Exploring Light and Sound

Core Knowledge®

ISBN: 978-1-68380-578-6

Exploring Light and Sound

Table of Contents

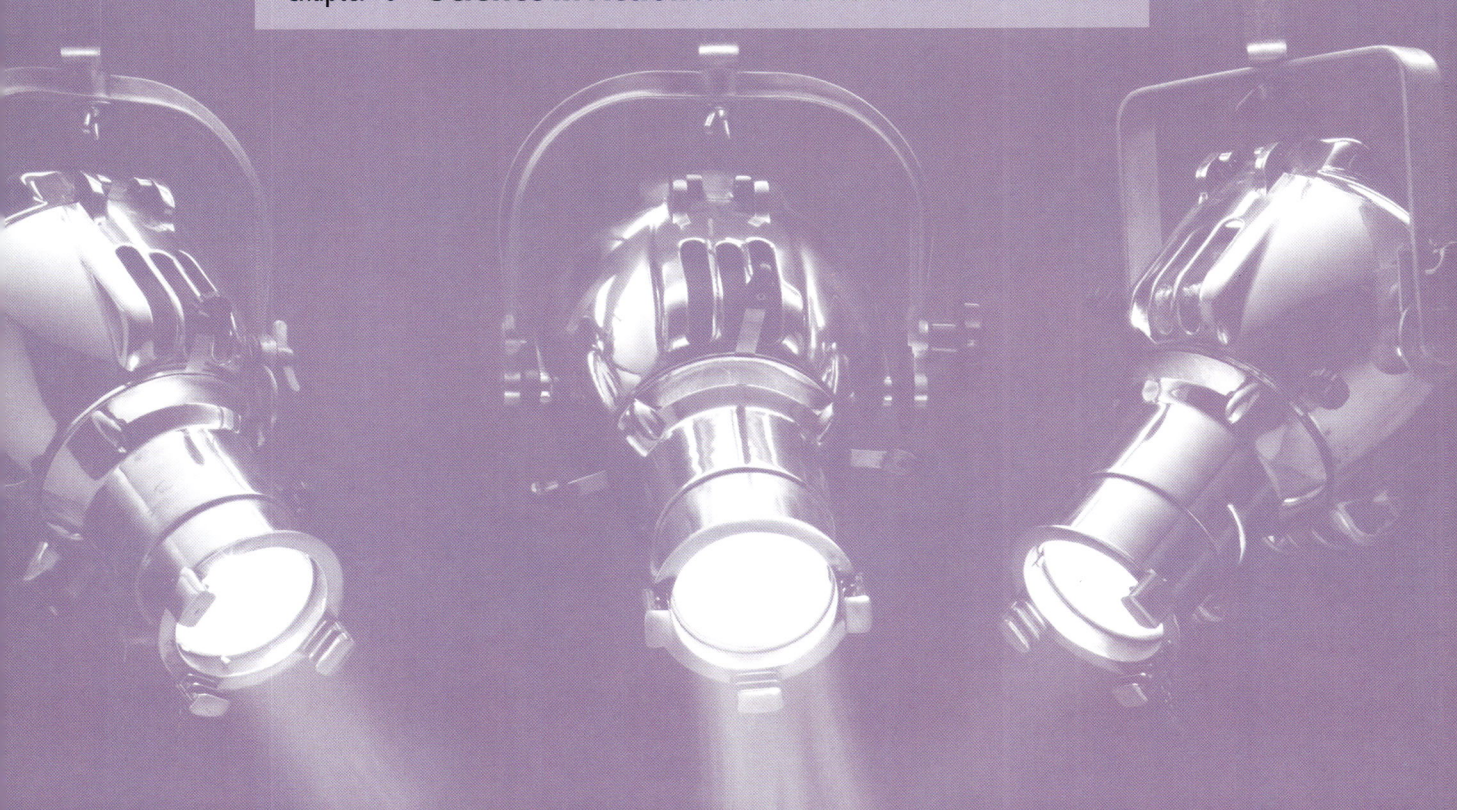

A Trip to a Musical Concert

Mr. Ruiz has big news! His class is going to a musical concert. The class is very excited. Their seats are in the last row of the theater. Mr. Ruiz says, "Don't worry. Everyone will be able to see and hear just fine."

The concert is amazing! There are lights. There are music and sound effects. On the ride back to school, the students talk about their favorite parts. They wonder how it is that everyone in the theater could see and hear so well. Mr. Ruiz says, "Let's investigate to find out!"

Sound and Vibration

A noisy jet passes overhead. The windows of the homes rattle. How are the two connected? Think about causes and effects. Yes, something about the sound of the jet causes the windows to rattle!

If you pluck a guitar string, it will vibrate. To vibrate means to move back and forth quickly. A moving string causes the air to vibrate. The moving air causes parts of your ears to vibrate. So, a vibrating string causes a sound you can hear with your ears.

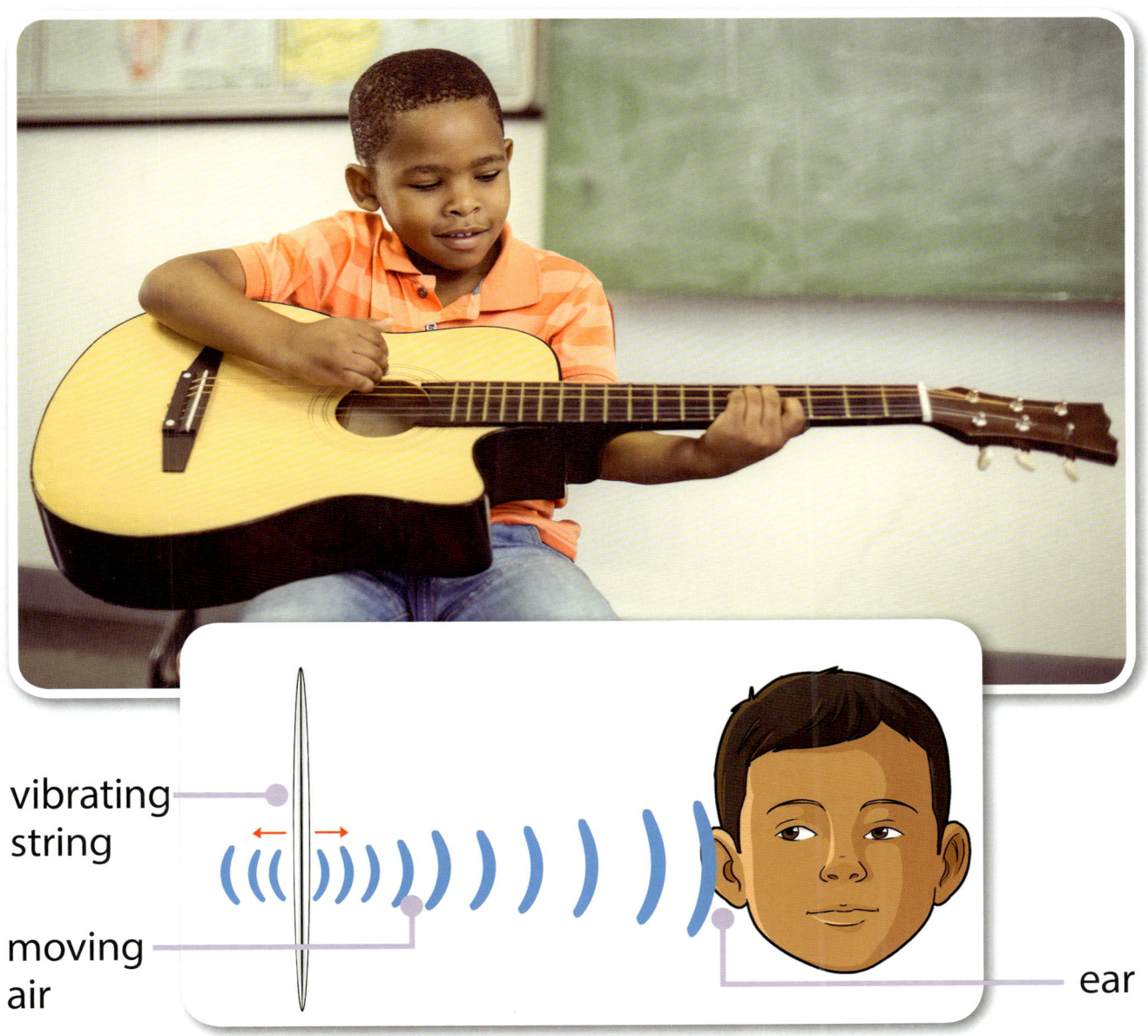

vibrating string

moving air

ear

Can you explain how the windows rattle? The jet engine vibrates. It makes air around the engine vibrate. The vibrating air causes the windows to rattle or vibrate.

Mr. Ruiz asks his class to think of things that vibrated at the concert. The instruments vibrated when played by the musicians. The microphones vibrated. The big speakers around the theater vibrated. The air around vibrated. Vibrations made the sounds that we heard.

There were many sounds during the concert. There was a cymbal clanging. There was a drum banging. There was a horn tooting. Each sound was made by something that vibrated. That made the air vibrate. How does this help you explain why people in the last row can hear the sounds?

Light

Mr. Ruiz tells his class a story about two brothers. One brother walks into the bedroom and sees his younger brother on his hands and knees under the lamp. The younger brother says he is looking for a pencil he lost. The older brother says, "Are you sure you lost it here by the lamp?" The younger brother says, "No, but the light is much better here." What's funny about this story?

Some objects around you give off light. The sun gives off light. Lamps give off light.

Nightlights give off light.

Lit candles give off light.

Mobile phones give off light. We can see light when it enters our eyes.

Some objects give off very bright light. A fire gives off bright light. It lights the area nearby.

The sun's light is so bright that it can hurt your eyes. Never look directly at the sun.

Other objects give off dim light. A glow stick gives off a dim light.

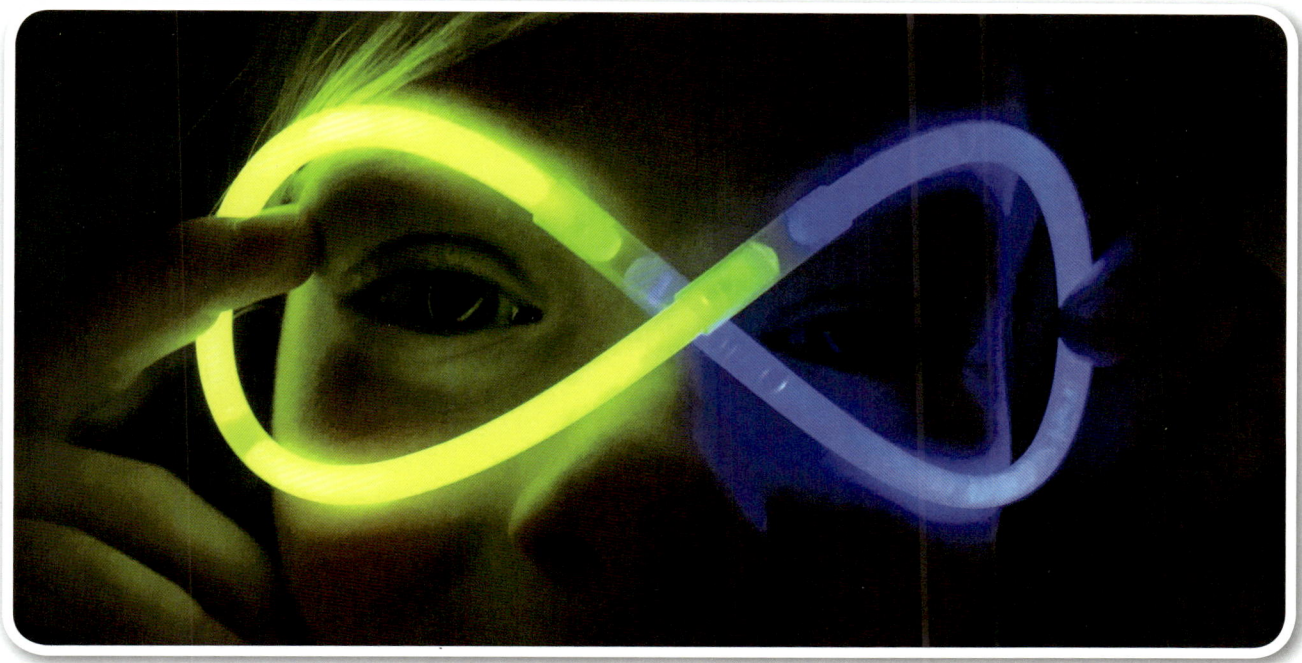

A flashlight also gives off light. Sometimes the light is bright. Sometimes the light is dim. Why might someone want to read by a bright light?

Many things do not give off light. Flowers, buildings, and fences do not give off light. People and dogs do not give off light either! You see these things only when light shines on them. When you are outdoors in the daytime, sunlight shines on these objects.

When you are indoors in the daytime, sunlight shines through windows or from light from lamps. At night, light from lamps shines on objects.

Think about opening the door to a dark closet just a bit. Without light, you cannot see what is in there. That can be mysterious!

But if you turn on a light, you can see inside the closet. It's just some shoes, some shirts, some dresses, and toys in there!

Some room lights have a dimmer switch. You can make the light dim, bright, or somewhere in between. It can be hard to read a book if the light is too dim or too bright. Using a dimmer switch can make the light perfect for reading.

Think again about the story of the brother looking for his pencil. Why was he under the light from the lamp? What could the older brother do to help find the pencil?

Light and Materials

"Remember when we went to the musical concert?" Mr. Ruiz asks. "We saw people and instruments in the bright spotlights. Look at this picture to help you remember. What did you also see under the drums?" The class yells all at once, "Shadows!"

How does the shadow in this picture happen? Light shines down on a person. Her body blocks the light. A darker shape appears behind her. That darker shape is called a shadow. Shadows are darker because there is less light there.

A shadow is not an object—like a ball, a bike, or a person. You need three things to make a shadow:

1. light shining

2. an object blocking the light

3. a surface on the other side of the object

Find the three things that make a shadow in this picture.

Many kinds of materials block light. All these materials can be used to make shadows. They all can block light.

This ball is made from rubber bands. The ball blocks light.

This bat is made of wood, and wood blocks light. The ball is made of many materials. The ball blocks light, too.

The stones are objects, and they block light.

These students want to learn more about light and shadows.
They are planning an investigation.
They have the three things needed to make a shadow:

1. a light

2. an object

3. a surface

What did the students do to make the shadow?

Other materials block only some light. They let some light pass through to the surface behind them. There may be a shadow, but it will not be as dark.

Balloons can be made from different materials. These balloons let some light through.

Some materials do not block much light. They let light pass through to the surfaces behind them. You can see well through materials that let all light pass through.

Water and clear glass let light pass through.

Air lets the light pass through. You can see a long way into the sky.

Light beams move in straight lines. The sun is shining down in this canyon. The light shines in a straight path.

Some objects can change the direction of a light beam, or reflect it. Materials that reflect light beams are often smooth and shiny.

Mirrors are good at reflecting light beams. Some mirrors are made of glass with a thin metal coating. Other mirrors are made of shiny metal.

When a mirror reflects a light beam, the light changes direction. It hits a surface in a different place.

Solving Problems with Light or Sound

Mr. Ruiz says to his students, "Do you remember that you were worried about sitting in the last row of the theater? How can people far away see and hear?"

The students have many ideas. They talk about a microphone placed near a musician. Maybe there are large speakers in the theater.

They also talk about lights that follow the action on stage. They are called spotlights.

How did these tools help the audience hear and see?

Now Mr. Ruiz has big news for his class. They will put on a show for their school! The class votes to tell the story of *The Little Red Hen* with music. First, they read the story. Then they talk about their questions. Who wants to sing? Who wants to play an instrument? Will there be lights and shadows?

Once upon a time
there was a little red hen

Next they plan their show. They want costumes and masks.
They want spotlights. They want sound effects.

The class also wants the audience to participate with sound. The audience will make animal sounds during parts of the show. They will say "Uh oh!" during parts of the show. But there is a problem. How will the players signal the audience to do the parts?

Students must solve the problem. They need to work together to figure out how to use light or sound to signal to the audience when to participate. The students use three main steps to solve the problem.

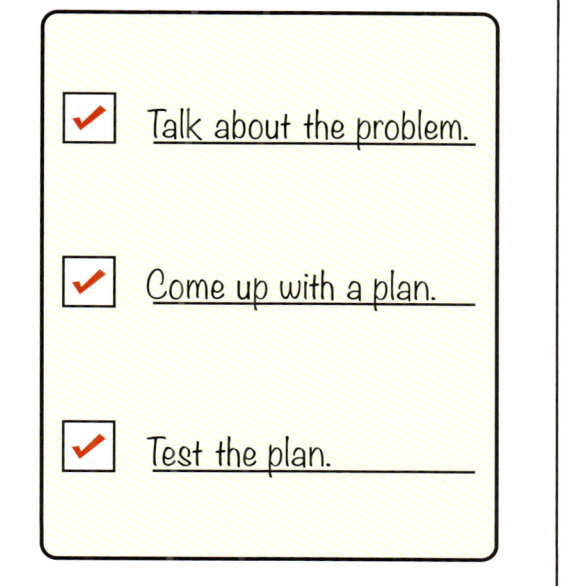

Now you know more about light and sound. How would you solve the problem?

How could you use light or sound to let audience members know when to make sound effects?

How could you let them know when to make a "meow" sound?

How could you let them know when to make a "woof" sound?

How could you let them know when to make a "cluck" sound?

How could you let them know when to make a "quack" sound?

Science in Action

Using Light as a Tool

Mr. Ruiz's class enjoyed using light and sound in their stage performance. His students used light and sound to communicate. But light and sound can be used for other things, too.

Mr. Ruiz wants his students to look for more ways that light is useful in unexpected places. They walk to a dentist's office near their school. The dentist shows them the tools that she uses to care for her patients' teeth. But how does a dentist use light as a tool?

The students are surprised to see that some of the dentist's tools are tiny, bright lights! One of the tools shines a purple light inside a patient's mouth. The light causes changes. It can whiten teeth and harden the fillings in cavities.

Another tool of dentists is a laser. A laser is a strong, narrow beam of light. It's very powerful. A laser beam can cut through the hard surface of a tooth. The dentist uses the laser tool to repair a tooth. The dentist says the laser is an easier tool to use than a drill.

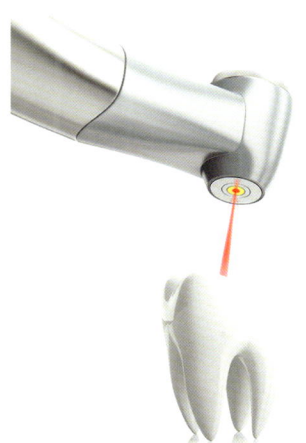

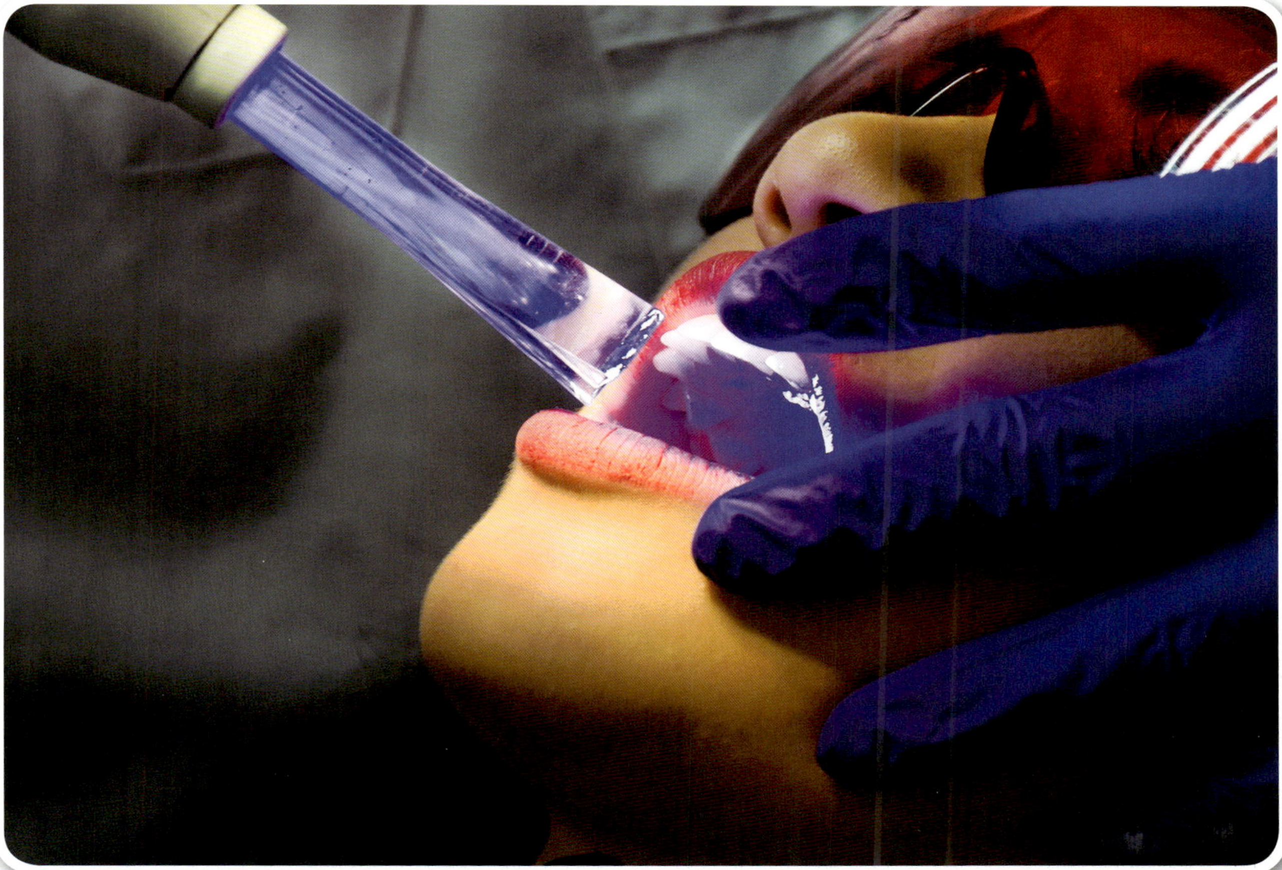

Back at school, Mr. Ruiz explains that lasers are used for other things, too. Strong lasers can be used to cut metal and other hard materials in factories that make things. Doctors performing surgery use lasers.

Mr. Ruiz says that a scientist named Gordon Gould invented the laser. Gould did what all scientists do—he started by asking a question. What would happen if a large amount of light energy was focused into one small beam? Asking a question is one of the first steps in the scientific method.

Gordon Gould

Gordon Gould was a scientist who studied matter and energy. He was especially interested in how light behaves. The questions that Gordon Gould asked about light led to the invention of the laser. This was one of the most important inventions of the last century because lasers can be used in so many ways. Lasers have made many processes faster and easier. They perform tasks over and over again without wearing out the way other tools do.

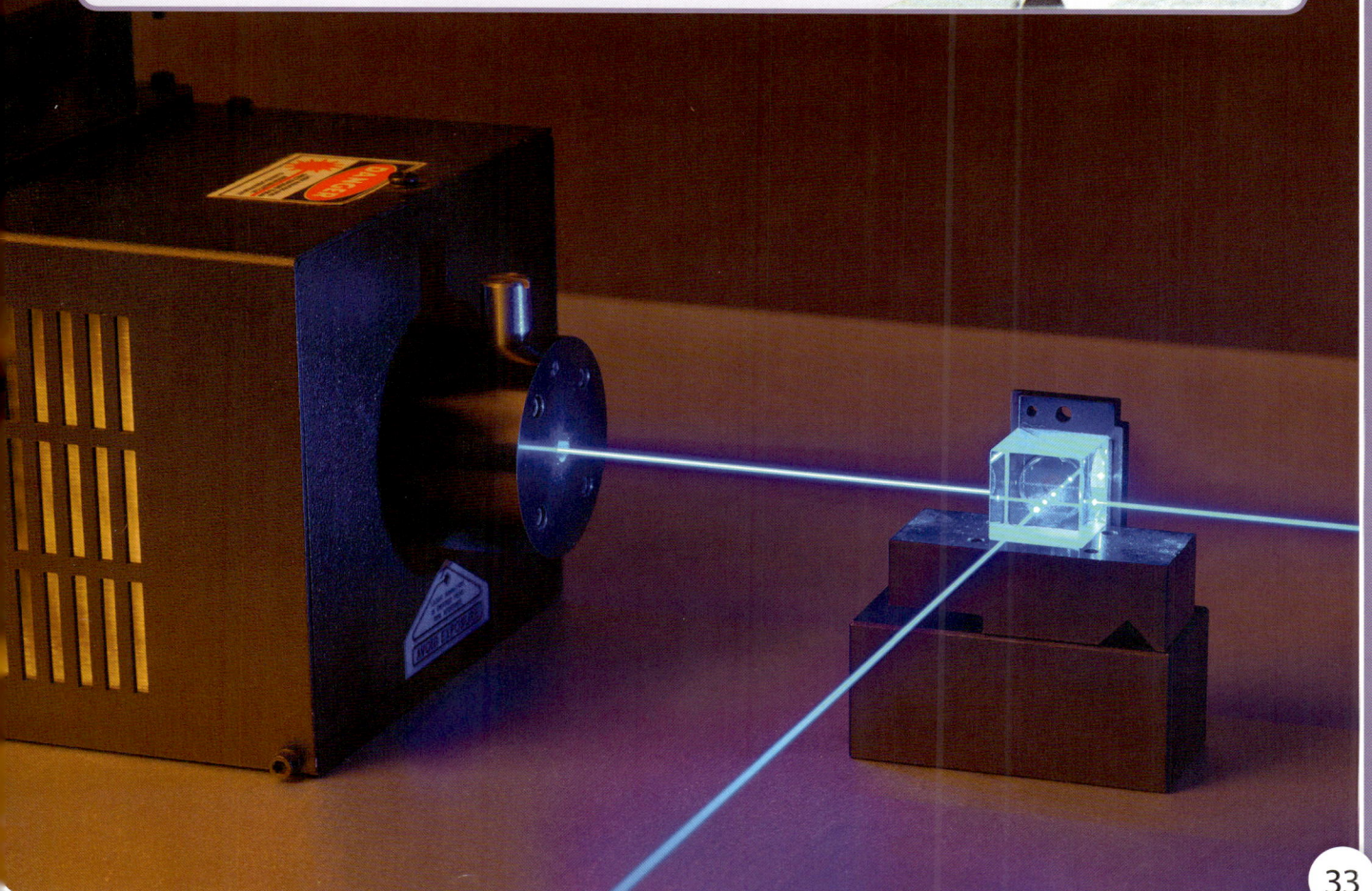

Core Knowledge®

CKSci™
Core Knowledge SCIENCE™

Series Editor-in-Chief
E.D. Hirsch Jr.

Editorial Directors
Daniel H. Franck and Richard B. Talbot

Subject Matter Expert

Martin Rosenberg, PhD
Teacher of Physics and Computer Science
SAR High School
Riverdale, New York

Illustration and Photo Credits